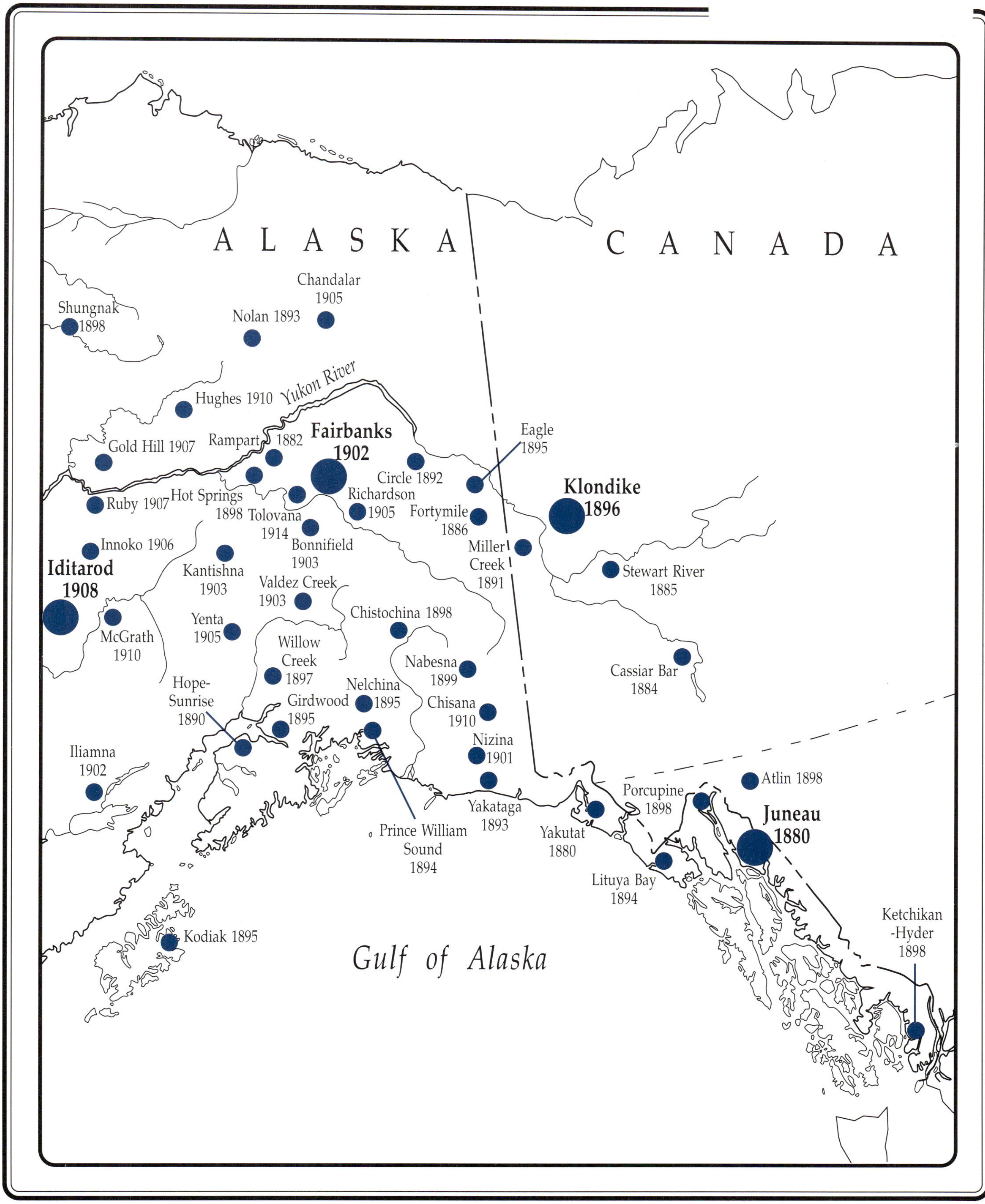
A L A S K A
C A N A D A
Chandalar 1905
Shungnak 1898
Nolan 1893
Hughes 1910
Yukon River
Gold Hill 1907
Rampart 1882
Fairbanks 1902
Eagle 1895
Circle 1892
Klondike 1896
Ruby 1907
Hot Springs 1898
Tolovana 1914
Richardson 1905
Fortymile 1886
Innoko 1906
Bonnifield 1903
Miller Creek 1891
Stewart River 1885
Iditarod 1908
Kantishna 1903
Valdez Creek 1903
McGrath 1910
Yenta 1905
Chistochina 1898
Willow Creek 1897
Nabesna 1899
Cassiar Bar 1884
Hope-Sunrise 1890
Nelchina 1895
Girdwood 1895
Chisana 1910
Nizina 1901
Iliamna 1902
Atlin 1898
Porcupine 1898
Yakataga 1893
Juneau 1880
Prince William Sound 1894
Yakutat 1880
Lituya Bay 1894
Ketchikan -Hyder 1898
Kodiak 1895
Gulf of Alaska

Gold rush stampeders building boats at Lake Bennett for the river journey to the Klondike, 1898 (University of Washington, Hegg 229).

ALASKA GOLD RUSH CENTENNIAL™

Gold Fever in the North
The Alaska-Yukon Gold Rush Era

A Gold Rush Centennial Exhibition

Dean Littlepage, Guest Curator

Anchorage Museum of History and Art
Municipality of Anchorage
1997

Catalog printing by First National Bank of Anchorage

Gold Fever in the North: The Alaska-Yukon Gold Rush Era / Dean Littlepage
ISBN 1-885267-01-0

This publication is a companion volume to the exhibition *Gold Fever in the North,* a gold rush centennial exhibit held May 8-November 9, 1997 at the Anchorage Museum.

121 West Seventh Avenue
Anchorage, Alaska 99501

Printed in Anchorage, Alaska

Curator: Dean Littlepage
Archivist: Diane Brenner
Exhibit design: David Nicholls
Exhibit installation: Allan Shayer, Ted Gardeline
Fundraising: Walter L. Hays
Registrar: Barbara Geib
Publication design: Ruth Ann Dickie

Cover illustrations

Front cover:

Upper left: Gold rush miners at work with sluice, shovel and pan (University of Alaska Fairbanks Library 73-50-13N).

Upper right: Starting out on the Chilkoot Trail, c. 1897 (Alaska State Library 87-664, Winter and Pond photograph).

Lower: Flat City, Alaska, in the Iditarod mining district. The towns of Iditarod and Flat were born in a 1909-10 gold rush. (Anchorage Museum B67.13.45, Ralph Lomen photograph).

Back cover:

A map of the Chilkoot (Dyea) and White Pass (Skaguay) Trails, first published in 1897. Skagway was spelled with a *u* before 1900 (W.B. Conkey Company, *En Route to the Klondike*; also published in the *Seattle Post-Intelligencer*).

Introduction

Spreading slowly into Southeast Alaska and along the Yukon River in the 1880s and early 1890s, northern gold fever became a national epidemic with the great Klondike gold rush of 1897-98. The Klondike stampede in turn triggered rushes to dozens of discoveries in Alaska, from Southeast to the Brooks Range and from the Wrangell Mountains to the Bering Sea.

The gold rush era was a time when ordinary people did extraordinary things. Driven by the economic depression of the 1890s, many untypical goldseekers — accountants, bank clerks, college students, storekeepers, and others — swelled the ranks of the stampeders. Many of them had never spent a single night in the outdoors until they joined the rush to the northern wilderness. Very few became rich, but most had one thing in common: the gold rush was the greatest adventure of their lives.

By the time the gold rush era faded in the 1910s, the stampeders had irrevocably changed Alaska and the Yukon, bringing towns, trails, railroads, and new ways of life to the North. They had shaped the settlement on the land, and the history and spirit of the North, for generations to come.

Special thanks to Candy Waugaman
for sharing her vast personal collection
and her passion for Alaska's past.

Acknowledgements

The generosity of our financial contributors, the enthusiastic support of public collection managers and private collectors, and the assistance of our advisors made *Gold Fever in the North* possible.

The generous support of First National Bank of Anchorage and Visible Ink made this publication possible.

Financial support

Leadership gifts:
Grand Circle Foundation
Anchorage Museum Foundation (Kreielsheimer Fund)
KTUU Television

Corporate sponsors:
AT&T Alascom
Anchorage Daily News
Alyeska Pipeline Service Company
The Chevron Companies
First National Bank of Anchorage
The Holland America Family of Companies
Kennecott Minerals Companies
Lynden Transport
MACtel Cellular Systems
Totem Ocean Trailer Express
Unocal

Additional support:
Aeromap U.S.
ATU Telecommunications
Carlile Enterprises Inc.
Guess & Rudd, P.C.
Placer Dome U.S.
State of Alaska, Division of Tourism and Office of History and Archaeology
Tesoro Alaska Petroleum Company
Visible Ink
WGM, Inc.

Patrons:
Abacus Minerals Corporation
Birch, Horton, Bittner, Inc.
Louise Gallop
Heller Ehrman White & McAuliffe
North Pacific Mining Company
Dave Parish & Associates
Silverado Mines, Ltd.
Cha Smith
Sumitomo Metal Mining Arizona, Inc.

Friends:
Coeur Alaska, Inc.
Cominco Alaska
Consolidated Nevada Goldfields Corp.
Hawley Resource Group
NANA-Dynatec JV
Ounalashka Corporation
Joseph Perkins and Rebecca Graham
Petro Marine Services
Yukon Equipment

Donors:
AMAX Gold, Inc.
Robert B. Atwood
Fairway Fast Freight
Mark and Barbara Fullmer
Johnson & Higgins
Turnagain Products

Objects and photographs
Alaska State Library, Juneau
Alaska Department of Transportation
Alaska State Museum, Juneau
Anchorage Museum of History and Art
Leroy J. Barker, Anchorage
Melvin and Mary Barry, Anchorage
Bicycle Museum of America, Orchard Park, New York
Phyllis Brown, Skagway (George and Edna Rapuzzi Collection)
Roger Brown, Brentwood, New Hampshire
Central Historical Society
The Chevron Companies
Keith Christenson, Eagle River
Chugach National Forest
Columbia River Maritime Museum, Astoria, Oregon
Crow Creek Mine, Girdwood
Fairbanks Gold Mining Inc.
C.C. Filson Company, Seattle
Friends of the Tanana Valley Railroad, Fairbanks
Louise Gallop, Anchorage
Glenbow Archives, Calgary, Alberta
Don Grybeck, Anchorage
Chuck Hawley, Anchorage
Miriam and Hilary Hilscher, Anchorage
Holland America/Westours
Carol Huber, Anchorage
Michael Jackson, Anchorage
Juneau-Douglas City Museum
Klondike Gold Rush National Historic Park, Seattle
Klondike Gold Rush National Historic Park, Skagway
Dean Littlepage, Eagle River
Z.J. Loussac Library, Anchorage
Museum of Alaska Transportation and Industry, Wasilla
Museum of History and Industry, Seattle
National Archives, Alaska Region
National Archives, Washington, D.C.
National Archives of Canada, Ottawa
National Park Service, Alaska Region
Northern Land Use Research, Inc., Fairbanks
Parks Canada, Dawson City, Yukon Territory
Pioneer Museum, Fairbanks
Pioneers of Alaska Auxiliary #4, Lillian S. Watson Collection
Puget Sound Maritime Historical Society, Seattle
Dave Schwegel, Fairbanks
Tim Sczawinski, Seward
Morgan Sherwood, Davis, California
Oscar Tweiten, Fairbanks
U.S. Geological Survey
University of Alaska Fairbanks Museum
University of Alaska Fairbanks Rasmuson Library
University of Washington Allen Library, Seattle
White Pass and Yukon Route, Skagway
Yukon Archives, Whitehorse, Yukon Territory

Advisors
Jo Antonson
Earl Beistline
Pete Bowers
Michael Gates
Don Grybeck
Logan Hovis
Ann Kain
Bill Mackie
Tom Morgan
Frank Norris
Becky Saleeby
Charlie Tryck
Gib Whitehead

Music
Jubilee
Bill Staines

Contents

A Klondike Supply List

Year's Supply and Outfit for
Two Men costs $300

—

If the prospector can afford it, he should add to this outfit one rifle, one shotgun, one gill net and a supply of ammunition, bringing the total cost of an outfit up to $350.

PROVISIONS	POUNDS
Flour	800
Bacon	400
Beans	300
Dried fruit (assorted)	50
Sugar (loaf)	50
Condensed milk	25
Coffee	40
Tea	10
Corn Meal	25
Rice	20
Baking Powder	10
Salt	15
Oatmeal	25
Evaporated potatoes	20
Evaporated onions	10
Pepper	2
Soap	25
Evaporated vinegar (4 pints)	
Matches (3 dozen boxes)	
Total	1827 lbs.

TENTS, BLANKETS AND CLOTHING

1 wall tent, 10x8 feet.
4 pair heavy woolen blankets.
1 pair rubber blankets.
1 dozen heavy woolen socks.
4 suits heavy woolen underwear.
4 heavy overshirts.
4 pair heavy overalls.
4 pair heavy hip boots, rubber.
4 pair shoes.
2 heavy woolen suits.
2 Mackinaw suits.
6 pair heavy woolen mittens.
4 pair heavy woolen gloves.
2 pair snow shoes.
2 pair snow glasses.
2 caps.
2 broad brimmed heavy hats.
Total weight about 200 pounds.

HARDWARE

1 Yukon stove
3 lengths stove pipe.
1 large frying pan.
1 baking pan.
1 bread pan, 10 quarts.
3 granite kettles, 10, 8, and 6 quarts.
1 coffee pot.
2 granite plates.
2 knives, forks and spoons.
2 granite cups.
1 large spoon.
1 3° pound ax.
1 small hand ax.
1 hand saw.
100 feet half inch rope.
Nails.
Pitch.
Oakum for boat.
2 picks.
2 shovels.

	POUNDS
Estimated weight hardware	100
Estimated weight clothing, etc.	200
Weight of provisions	1827
Grand total weight	2127

Source: Millroy's Map, 1897 (Anchorage Museum Archives Collection).

Klondike supplies on a Seattle street, 1897 (Museum of History and Industry 401).

Miners at the mouth of the Fortymile River, 1895 (Veazie Wilson, *Glimpses of Alaska,* Parks Canada Collection).

The Early Years

Everybody is looking for gold, some finding it and some getting nothing, a few becoming rich, but the greater number only making a living, and all working very, very hard.

Father William Judge at Fortymile, 1894

In the years before the discovery of gold in the Klondike, gold fever spread slowly north through British Columbia and Southeast Alaska into the Yukon River basin. Discoveries in the Cassiar Mountains of British Columbia in the 1860s and 1870s put Wrangell, in Southeast Alaska, on the map as the gateway to gold country. The first prospectors arrived on the Yukon in 1873, following the long, overland Hudson's Bay Company route from the east.

In 1880, two events took place that ushered in the northern gold rush era: the discovery of the Juneau Gold Belt, and the opening of the Chilkoot

The Treadwell Mine in Douglas, across Gastineau Channel from Juneau. Claims John Treadwell bought here in 1881 for $400 yielded $67 million in gold (University of Alaska Fairbanks Library 82-70-5).

Trail. Both involved Southeast Alaska's Tlingit Native people.

George Pilz, a mining engineer operating out of Sitka, bankrolled pairs of prospectors in his search for Southeast gold. One pair, Joe Juneau and Richard Harris, spent most of their time, according to Pilz, on "hooch and squaws," but got a helping hand from a local Tlingit. Kowee, an Auke Bay Tlingit leader, led the two to their 1880 discovery on Gold Creek, near what is now downtown Juneau, and miners flocked to the Juneau Gold Belt.

At the head of the Inside Passage, the Chilkat band of the Tlingit controlled a trading trail that ran from the sea through the Coast Mountains to the Yukon River. The trail, over Chilkoot Pass, was the most practical route to the Yukon, and prospectors itched to hike through the Tlingit stronghold. In 1880, the Tlingit allowed the first group of miners, backed by a U.S. Navy detachment, to cross the Chilkoot.

Mining the Yukon

In 1886, two prospectors named Howard Franklin and Henry Madison found the first coarse gold in the Yukon basin, on the Fortymile River near the unsurveyed U.S./Canada border. A Native boy named Bob, relaying word of the strike to traders near Juneau, ran his hand through a sack of beans and told them, "Gold

Jack McQuesten, Father of the Yukon

Early Yukon River prospectors bought their supplies from Jack McQuesten, the Yukon trader who became known as the Father of the Yukon. McQuesten grubstaked prospectors on credit, in sharp contrast to his competitor John J. Healy, who insisted on being paid promptly for every last can of condensed milk he sold.

McQuesten knew the hardships of the country. He arrived on the Yukon in 1873, after two years of traveling through the North. During that time, he lived without so much as a bite of bread or flour, the usual staple of white men on the frontier.

He moved his trading posts with the tide of the miners. By the time of the rush to Birch Creek and Circle, he was the country's senior trader, and he became Circle City's leading citizen. McQuesten and other pioneer traders like Arthur Harper and Al Mayo made it possible for Yukon gold-seekers to fan out through the hills and find the gold that led to the founding of Forty Mile and Circle, and later, Dawson and Fairbanks.

Jack McQuesten's store in Circle (Yukon Archives 2443).

all same like this." With word of the discovery out, several hundred men stampeded to the Fortymile the next year. Soon there was a town, Forty Mile, where the river met the Yukon.

The early Fortymile prospectors and miners did not live in luxury on the Yukon; it would be several years before steamboats made frequent, reliable supply runs up the river. Men with nicknames like Cannibal Ike, Crooked-Leg Louie, Pete the Pig, and Tin Kettle George survived on moldy flour and yellowed bacon, and what game and fish they could muster. They boiled down moose and caribou bones and antlers to make a butter substitute, and brewed liquor out of boot soles.

As for readin' and litrachure and all that, wy, dammit, wen the first grub comes in the spring, we 'ave a meetin' an' we call all the boys together an' we app'int a chairman, an' then some one reads from the directions on the bakin'-powder boxes.

A Fortymile miner quoted by Josiah Spurr, 1897

In 1892, two Russian-Native prospectors named Pitka Pavaloff and Sergei Cherosky set out from the Fortymile, followed by a small army of white prospectors who thought the pair was on to something big. They were. Once they found the spot, on Birch Creek, the white prospectors pushed the discoverers aside and staked their claims.

The stampede that followed the Birch Creek discovery drained the Fortymile of its population and created Circle City. Circle's thousand residents bragged that the town was the "largest log-cabin city in the world."

Still, prospectors continued to search the lonely hills and streams for what they hoped would be the big strike, the one that would put the North on the map and make them rich.

The "Gay" Nineties and the Klondike strike

For many in the U.S., the Gay Nineties were not so gay. European immigrants flooded into the country, looking for opportunity and finding nothing. The Panic of 1893, a run on gold that depleted the U.S. Treasury, had mired the country in a deep economic depression. Gold *was* wealth and prosperity, both in the economy and in the national mind. The nation's mood was restless and root-

In this turn-of-the-century political cartoon, the Klondike is depicted as "Nugget Gulch" (Anchorage Museum, Sherwood Collection, Chris Arend photograph).

George Carmack, who filed the discovery claim on Bonanza Creek in the Klondike. His Native brother-in-law, Skookum Jim Mason, made the discovery (Alaska State Library 01-1923).

less, and all it needed to explode was a spark from the Klondike.

Up in the Yukon country, late in the summer of 1896, Skookum Jim Mason tracked a moose along Rabbit Creek, a tributary of the Trondiuck River. Mason, a Tagish Native, kneeled down to the water and saw gold in the sand. He and his partners, including his white brother-in-law, George Carmack, decided to stake claims. Carmack took the discovery claim, arguing that other whites would not allow a Native the honor.

The creek soon had a new name, Bonanza Creek, as did the Trondiuck, which the prospectors called the Klondike. Fortymile and Circle miners rushed there to stake. The following summer two steamships, the *Portland* and the *Excelsior*, sailed from the mouth of the Yukon to the West Coast with the first of the Klondike's gold.

Nearly 5,000 people met the *Portland* when it landed in Seattle on July 17, 1897, carrying two tons of gold. The news from the Klondike set off a wild stampede that would become the last great North American gold rush.

GOLD! GOLD! GOLD! GOLD!

Sixty-Eight Rich Men on the Steamer Portland.

STACKS OF YELLOW METAL!

Some Have $5,000, Many Have More, and a Few Bring Out $100,000 Each.

THE STEAMER CARRIES $700,000.

The Seattle Post-Intelligencer headlines on July 17, 1897, the day the Portland *landed in Seattle with two tons of gold from the Klondike* (Anchorage Museum Archives Collection).

The S.S. Queen *sailing from Seattle for the Klondike* (Museum of History and Industry 88.33.16, Anders Wilse photograph).

To the Klondike!

It was sad to contemplate this eager crowd of men from all classes of life, who ... hurried on in a mad rush for something they could never grasp.

John Sydney Webb, 1898

Pandemonium ruled the West Coast after the landing of the *Portland*. Seattle and other West Coast cities - San Francisco, Portland, Tacoma, Victoria and Vancouver - boomed as thousands of would-be prospectors set off for the Klondike. Outfitters hawked maps, guidebooks and catalogs of supplies. The many novices among the stampeders were fair game for salesmen of "necessities" for the Klondike.

Hastily-organized transportation companies leaped to life. Mariners revived practically any boat that would float for hauling the stampeders north. Even Seattle's mayor, W.D. Wood, joined the stampede — he resigned to set up a steamship and riverboat service to the Klondike.

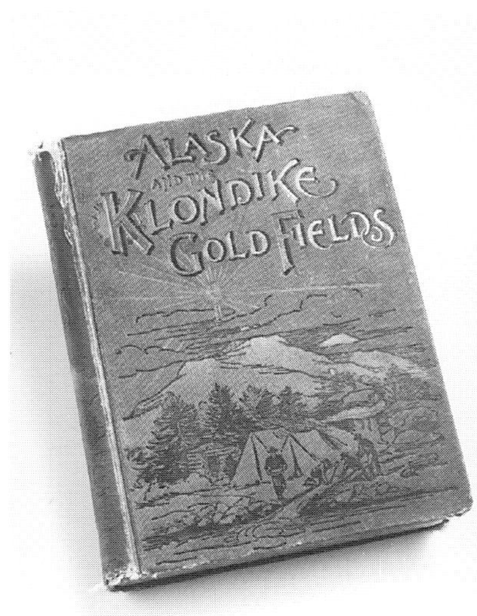

A gold rush guidebook by "the well-known author and traveler" A. C. Harris. Klondike guidebooks were not known for their accuracy (Anchorage Museum Archives Collection).

All that anyone hears at present is "Klondyke,"... it is impossible to escape it.

Seattle Times, July 24, 1897

Routes to the Klondike

There were really only two feasible choices for the journey to the Klondike: the routes over the mountain passes at the head of Southeast Alaska's Inside Passage, and the 4,500-mile "all-water" route across the Bering Sea and up the Yukon River.

The steamboat Alice *trapped in the ice on the Yukon River, 1897. Most of the stampeders who tried the "all-water" route spent the winter frozen in along the river* (Anchorage Museum B74.1.2).

Promoters publicized several other, much longer and more difficult overland routes, including "trails" through Valdez, Alaska; Ashcroft, British Columbia; and Edmonton, Alberta. Few stampeders who tried these difficult routes reached Dawson.

"The All-Canadian" route through Edmonton was a nightmare. Advertised as a quick six-week trip, some people took two years to reach Dawson, and most who tried turned back after sampling the non-existent "trail." At least 70 people died trying to reach the Klondike from Edmonton.

The most popular routes were the passes between the Inside Passage and the upper Yukon. About 40,000 to 50,000 people tried these two trails, the Chilkoot and White Pass Trails, in 1897-98. Dyea, at the foot of the Chilkoot, and Skagway, the jumping-off point for White Pass, exploded from tiny settlements to "cities" of thousands of transients as the stampede worked its way north.

I have stumbled upon a few tough corners of the globe during my wanderings. . ., but I think the most outrageously lawless quarter I ever struck was Skagway.

Alexander Macdonald in
Skagway, 1897

In Skagway, Soapy Smith and his gang of con men and thieves preyed on the thousand stampeders a month who flowed through the town. They fleeced the newcomers with a fake telegraph scam, rigged gambling, and outright robbery.

The majority of the Klondike stampeders chose the Chilkoot Trail, a route that had been well known among northern miners since its 1880 opening.

A pack train leaving Skagway for White Pass and the Yukon in 1897. So many horses died on the White Pass Trail that it came to be called the Dead Horse Trail (Anchorage Museum B64.1.31).

The Chilkoot Trail

All Fool's Day, and most all the fools are here.
Louie B. May on the Chilkoot Trail, April 1, 1898

A cart used to carry supplies on the lower part of the Chilkoot Trail during the Klondike Gold Rush (Anchorage Museum 94.44.1).

The uninhabited, snowbound peaks of the Coast Mountains rang with voices throughout the winter of 1897-98 as the Klondike-bound adventurers climbed toward Chilkoot Pass. The Chilkoot Trail, just 33 miles long, led them from the port at Dyea across the pass to Lakes Lindeman and Bennett, great glacially-carved lakes that feed the Yukon River. At the lakes, the stampeders could build boats, and after the ice went out, float the 550 miles down the river to the Klondike.

The stampeders had to struggle up to Chilkoot Pass with more than 1,000 pounds of food and gear apiece, all the supplies they would need to survive a year of prospecting in the Yukon. They might walk several hundred miles back and forth on the trail, carrying and caching successive loads as they moved slowly toward the pass.

At least four weeks on the trail, part of the winter camping and building boats at one of the frozen lakes, and two to three weeks on the river after the ice melted in May lay ahead for the would-be prospectors. After all that, they would reach

Nearing the summit of the Chilkoot Trail, 1898 (University of Alaska Fairbanks Library 78-111-7N).

the new town of Dawson, where the Klondike River met the Yukon, and gold lay waiting for the taking.

In single file we swing along
and soon we strike the stride,
That climbs the trail that stands on end
against the mountain side.
Stampeder poem, 1898

To reach the top of the pass, the stampeders climbed the Golden Stairs, more than a thousand snow steps carved into the mountainside. At the summit, the North-West Mounted Police checked each person for the required supplies and collected customs duties. Mountains of supplies lay cached in the pass as the stampeders went back for more of their gear. The savvy ones marked their caches with long poles; otherwise the frequent blizzards would soon cover them without a trace.

From Chilkoot Pass, it was still a journey, but at least a downhill one, to Lakes Lindeman and Bennett. The stampeders built makeshift cities on the shores of both lakes; Lindeman City even elected a town council that winter. Bennett swelled with stampeders who had traveled over White Pass from Skagway.

Sheep Camp, one of the transient villages that materialized along the Chilkoot Trail. The last refuge in the trees below the alpine climb to Chilkoot Pass, Sheep Camp had over 50 businesses by the spring of 1898. It was deserted by 1899 (Anchorage Museum B64.1.37).

Avalanche!

Just before noon on Palm Sunday, April 3, 1898, a huge snowslide between Sheep Camp and Chilkoot Pass buried scores of stampeders and killed about 60. Snow had fallen almost continuously in March, and as April began, warm winds combined with the heavy snow created ideal avalanche conditions. Native packers refused to work in the high country because of the danger, but many of the stampeders didn't heed this warning and continued plodding forward with their supplies.

Finally, a series of small avalanches convinced them how real the danger was. A large group tried to descend to safety on the trail, but the big slide caught them before they could reach safety at Sheep Camp.

Recovering a victim of the Palm Sunday avalanche (University of Alaska Fairbanks Library 67-17-20N).

The real work of the winter was boat building. The stampeders felled trees, sawed them into lumber, and nailed and caulked the lumber into boats for the trip downriver to Dawson. By spring, they had denuded the forest far back from the lakes, and had built over 7,000 boats.

The North-West Mounted Police numbered and registered each boat and each passenger, and set up checkpoints along the river to help the stampeders reach Dawson safely. In late May, when the ice melted off the lakes, the flotilla pushed off for the Klondike.

Down the Yukon to Dawson

The stampeders' trip from Bennett to Dawson was mainly a flat-water float, but wind-blown lakes and rapids lay ahead as dangers. Miles Canyon, Squaw Rapids, and Whitehorse Rapids, all in a brief stretch of the river, formed the biggest obstacle. A number of stampeders wrecked in the rapids or in the winds on the headwater lakes, and a few drowned.

More split up their partnerships on the way from the tensions built up over the long winter. At the Big Salmon River, ten men ended their partnership by ripping apart their boat, dividing it ten ways, and building ten new boats for themselves.

Finally the stampeders reached the Klondike River and landed at the hovel of tents and new buildings that was Dawson. It was a rude awakening. The good gold claims, it turned out, had been staked before they had even left the West Coast, but the Yukon was so isolated and communication so poor that the hordes outside had no way of knowing. The miners and prospectors who were already in the Yukon and Alaska had rushed to the Klondike and beaten the newcomers to the riches.

Disappointed, many returned south, but many hung on until news of gold discoveries in Alaska led them to stampede there — to Nome in 1899, Fairbanks in 1903-04, and to other discoveries in the years that followed. The overflow from the Klondike was headed for Alaska.

Leaving Lake Bennett for the Klondike (Anchorage Museum B64.1.43).

Everybody is crazy about Cape Nome. How I wished we had gone there instead of coming up here. . . .

Emily Craig in Dawson, 1899

A crowd of stampeders on the banks of the Yukon at Dawson, 1898 (Alaska State Library 34-80).

Cleaning up gold from a sluice (University of Alaska Fairbanks Library 89-166-05N).

Into Alaska

Several feet of the flume bottom was a thin sheet of solid gold particles of all sizes. Standing there on that gold carpet was a thrill you'd never forget, and when you reached down and let a handful run through your fingers — Alaska had you.

Kenneth Gideon at Crow Creek Mine, 1913

Disillusioned Klondikers and fresh recruits from the States headed for Alaska in droves. They came to Nome from Dawson in 1899 and 1900 by foot, dog team, riverboat, and even bicycle. With thousands more who came from the south by steamship, they built a tent city that stretched for 20 miles along the beach at Nome. By 1904, about half of Dawson's dwindling population had rushed to the new diggings at Fairbanks.

Besides the great fields at Nome and Fairbanks, many other gold districts across the territory kept Alaska's miners busy through the first decade of the 1900s. Then the last great discovery, at Otter Creek in 1908, brought thousands of stampeders to the remote Iditarod region. Smaller rushes to places like Chisana and Livengood kept the stampede era alive into the mid-1910s.

Frontier life and independent mining at each new gold camp lasted only a year or two. The trappings of civilization and industrial, corporate mining followed quickly. Independents, though, could go on to the newest discovery, and they did, until the names of gold camps were sprinkled all across the map of Alaska. Finally, when most of the rich deposits had been discovered, and World War I siphoned off the ranks of younger adventurers, the era of the independent gold stampeder came to a close.

Nome: Gold in the Ruby Sands

In 1898, out on the treeless west end of the Seward Peninsula, a howling Bering Sea storm forced four men to land as they sailed a schooner along the windswept shoreline. They explored up a nameless creek as the storm raged out on the sea, and one of them, John Brynteson, found some color. He soon returned to the creek with two different partners, and they found enough gold to stake claims. The "three lucky Swedes" (one was Norwegian) had discovered Anvil Creek, one of the richest gold creeks in Alaska.

After word of the strike got out, miners from the nearby Council district and luckless prospectors from the Klondike arrived and began an orgy of claim staking and jumping. Mostly Americans, they resented the success of the "foreigners" and overstaked the discoverers' claims. Speculators soon blanketed the rest of the area with claims.

When thousands more stampeders arrived in 1899 and 1900, there were few people working and almost no nearby ground unclaimed. Then, by another lucky accident, an Idaho miner named John Hummel, too weak with scurvy to travel far from the beach, found gold in the ruby sands on the shore of the Bering Sea. Soon prospectors lined the beach, a shovel's length apart, cleaning up gold with pan and rocker. The law didn't allow claim staking on the beach, so it became "a poor man's paradise."

Mining the Nome beach. A Nome newspaper complained that with all the miners' homemade equipment, the beach looked like "the world's longest junkyard" (Anchorage Museum B64.1.108).

A July 1900 traffic jam in Nome. Only two years before, what was now Front Street had been a wilderness (Anchorage Museum B72.46.110).

When miners found old, buried beach lines inland, the deposits turned out to be some of Alaska's richest gold diggings, rich enough that over time Nome has proved to be Alaska's third-richest gold district, behind only Fairbanks and Juneau.

Fairbanks: the Alaskan Klondike

It was 1901 on the Chena River, and E.T. Barnette was anything but happy. Shallows on the river had forced the captain of the riverboat *Lavelle Young* to unceremoniously dump Barnette and his tons of supplies on the river bank, several hundred miles short of where he had planned to set up a trading post.

A prospector named Felix Pedro, though, had spied steam from the riverboat as he perched on a hilltop several miles away. Pedro, who was on a lonely search for gold in the Tanana Hills, made his way to the river to meet Barnette. Pedro stocked up on supplies and prospected in the hills that winter instead of retreating to Circle, a 150-mile walk north across the hills. It paid off; Pedro found gold, and Barnette's trading post became Fairbanks.

A tremendously rich strike has been made 18 miles north of the Tanana River, in American territory . . . and a wild stampede is in progress.
Yukon Sun, January 17, 1903

After Pedro's discovery, Barnette sent his Japanese employee, seasoned dog musher Jujiro Wada, to Dawson to paint a glowing picture of "a second Klondike." Wada's report to the *Yukon Sun* sent several hundred Dawsonites on a 50-below-zero stampede to the Chena in January, 1903. When they reached Fairbanks, they found that speculators had tied up all the good claims. The newcomers rebelled, threatening to lynch first Wada, and then Barnette.

The standoff ended peaceably, and over the next few years, Fairbanks did become a second Klondike. The rolling hills and streams looked like the Klondike, and like the Klondike, the creeks were rich in gold — so rich that Fairbanks became the richest gold district in Alaska.

Iditarod: Up the Lazy River

On Christmas Day, 1908, John Beaton and William Dikeman got a surprise gift: gold in their first prospect hole on Otter Creek, on the upper Iditarod River. The two prospectors had been less than wildly successful so far in the gold rush — they had both rushed to the Klondike, Nome and Fairbanks without much to show for it. Their new partnership, though, was looking promising.

A model of the steamboat Lavelle Young. *Its grounding on the Chena River led to the founding of Fairbanks.* (Anchorage Museum 97.4, Chris Arend photograph).

The Iditarod River below Dykeman is a sluggish, deep tea-colored canal-like stream and is, without a doubt, the crookedest river in Alaska.
Anton Eide, 1910

When the rivers thawed in the spring of 1909, they piloted their small sternwheeler, the *K.P.M.*, over 500 miles on the slow, winding Iditarod and Innoko Rivers to Ophir, only 90 miles overland from their discovery, to record their claims. Word spread to Fairbanks and to prospectors closer by, on the Innoko and Kuskokwim Rivers.

The Innoko and Kuskokwim prospectors reached the Iditarod that

A cavalcade of miners, bringing in a quarter million dollars in gold from Cleary Creek, near Fairbanks, 1905. Only four years after Felix Pedro discovered gold, the Fairbanks district was producing nearly half the gold mined in Alaska (Anchorage Museum B64.1.68).

summer, but most of the stampeders from Fairbanks, caught by low water and an early freezeup, had to winter short of the new diggings. Supplies were scarce in the remote Iditarod country, and little mining took place in the summer of 1909.

The following summer the riverboats *Tanana* and *Delta* steamed as far as possible up the shallow river, and in a flat, treeless bog on a muddy riverbank, the boats' passengers founded the new city of Iditarod. Iditarod was the supply center for the new gold district, which was still farther upstream. The town had no gold, but its entrepreneurs managed to squeeze plenty of gold out of the miners on Otter Creek.

Sunrise, a gold rush town founded in 1896 on the northern Kenai Peninsula. Sunrise's fortunes had faded by 1918 (U.S. Geological Survey, Moffitt 196).

Cook Inlet: Mountains and Gold

From the trading post at Kenai in 1888, a mysterious character named Al King rowed up Turnagain Arm on a search for gold in the Kenai Mountains. King returned to Kenai two summers later with two pokes of gold, paid his bills, and moved on, leaving rumors behind him. Actually, King had only *rediscovered* gold on the Kenai Peninsula. Russian engineer Peter Doroshin had found the first gold in Alaska on the Kenai in 1849.

Others took up the chase, scoring discoveries in the mid-1890s that would soon spawn the towns of Hope, Sunrise and Girdwood.

A group of prospectors landed at Resurrection Creek in 1895 and called the spot Hope, as the story goes, after Percy Hope, the youngest in the party. In 1896, the same year Skookum Jim and George Carmack staked their claims in the Klondike, a small stampede engulfed Hope and Sunrise.

After the turn of the century, prospectors hit paydirt on the Yentna and upper Susitna Rivers, bringing more stampeders to upper Cook Inlet. While none of the Cook Inlet districts made the ranks of the top Alaska gold producers, the search for gold brought white settlement to the area that is now home to more than half the state's population.

The riverboats Tanana *and* Delta, *landing at the site of Iditarod on June 1, 1910 (*Museum of History and Industry 12878*).*

The Bergman String Band enjoying gold rush life on the Koyukuk River, 1899 (Anchorage Museum B89.24.183, J. N. Wyman photograph).

Life in the North

Finally the ball was over . . . . How strange we all must have looked — five hundred people in correctly formal clothes, trudging through the snows to our homes against a backdrop of shacks and log cabins and the dark bulk of the lonely hills.

Laura Berton in Dawson, 1907

Stampeders who joined the rush to the North left friends, family, and familiar places far behind. Distance and ice isolated them for as much as eight months a year, so they built their own society. They frequented saloons, formed book clubs and bands, crowded into halls to see pioneer theater groups, and threw formal parties on the flimsiest of excuses.

Reading the first news of the Spanish-American War in Dawson, 1898. The stampeders crowded streets and halls to hear readings whenever someone brought in a newspaper, no matter how dated (University of Alaska Fairbanks 70-58-434).

A prospector's life

The early prospectors had to be a hardy bunch, carrying heavy loads of supplies and equipment as they hiked the creeks looking for gold. They built small, low-profile, dirt-floor cabins, and furnished them with just the basics: a handmade table, chairs, bunks, shelves and a wood-burning stove. Only a red-hot stove could keep even the small cabins warm when the winter temperatures plunged far below zero.

The prospectors' basic foods were the "three B's" — bread, beans and bacon. They baked bread from flour and sourdough leavening, which they made by fermenting a mixture of flour, sugar and water. Practically every prospector's cabin had a pail of sourdough by the stove.

Entertainment

The social life of the early mining camps revolved around saloons, which were usually dance and gambling halls as well. There were dozens of saloons in places like Nome, Skagway and Dawson. One Dawson saloon advertised that the house ran "crap, chuck, and draw poker, blackjack, roulette wheels and faro banks," and that the bar had every known fluid except water.

(There were) Scotch ballads and songs of the Russian steppes, Highland flings and sword dances, Cossack dances, songs and dances of the Scandinavian countries, Italy, Greece, Montenegro, France, Belgium, Irish jigs and reels, French Canadian, even a few American songs.

Harold Peckenpaugh at a tavern in the Iditarod District, 1913

Music, dances and theater made slightly more genteel amusement. Only six months after Iditarod was just a muddy bog on a riverbank, the townspeople held their first formal ball. The George Snow family brought theater to the

Wool coats like this "Mackinaw cruiser" were on every prospector's supply list. The manufacturer, C.C. Filson Company of Seattle, got its start during the Klondike rush. Filson began making this pattern of coat during the gold rush era, and still sells it (C.C. Filson Company, Chris Arend photograph).

Immigrants Among the Stampeders

In 1900, nearly half of Alaska's gold miners were foreign born. Many of them had been U.S. residents only briefly before the gold rush. The largest groups of foreign-born stampeders were Norwegian, Swedish, and British-Irish. There were no Chinese or Japanese in the mines, but they did other work in the camps. Chinese made up 5% of Alaska's population in 1900.

A decade later, gold strikes at Iditarod and Ruby in interior Alaska attracted southern Europeans. In Iditarod, Episcopal minister Hudson Stuck gave a fundraising speech for a new hospital to a group of miners, but got no reaction from the crowd. Only afterward did their supervisor tell Stuck the miners were all from Montenegro, and not a one spoke English.

Maori (New Zealand Native) men at their cabin near Dawson (Yukon Archives 686, E.A. Hegg photograph).

Fortymile and Circle, becoming the first white family to cross Chilkoot Pass in the process.

The stampeders held all kinds of sporting events — foot races, sled dog races, pole boat races, wilderness marathons and more. In Iditarod, a snowshoer beat a dog team in a 38-mile race in heavy snow. On the Koyukuk, the townspeople of Bergman built an outdoor boxing ring for what they billed as the first prize fight in the "frigid zone."

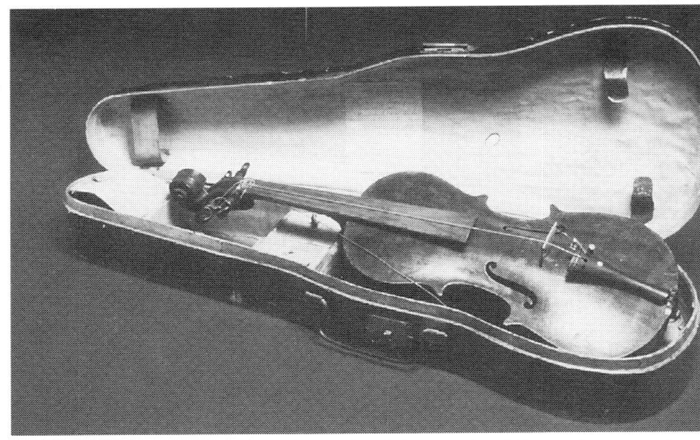

Nome miner Abe Ellis's violin, which he crafted from the wreckage of the freight barge Skookum *after a Bering Sea storm in 1901* (Alaska State Museum II-0-140, Mark Daughhetee photograph).

Women in the rush

Women did join the gold rush, but there were usually very few women in the new camps. The Mounties told Martha Black, who hiked the Chilkoot Trail in 1898, that she was the 631st woman past a checkpoint on Tagish Lake, compared to 18,000 men. On Cook Inlet, there were about a dozen couples and fifty single men in Sunrise in 1898, two years after the first rush. Women came to the North as entrepreneurs, seamstresses, entertainers, and homemakers.

A slot machine used in a gambling establishment in Skagway, c. 1900 (George and Edna Rapuzzi Collection, Mark Daughetee photograph).

The "dance hall girls," women like Klondike Kate Rockwell and Gussie Lamore, excited a lot of attention. Romance as spectacle reached new heights when Swiftwater Bill Gates greeted Lamore with a $50,000 gift and a proposal when she arrived in Dawson.

Prostitutes set up shop on the Line at Fairbanks, the Row at Iditarod, and the White Chapel outside Dawson. In Iditarod, the women of the Row held a town dance every year called the Pretzel Dance. They mixed with polite society at the town's masquerade balls, but obeying an unwritten rule, always left before the midnight unmasking.

The Yukon Order of Pioneers, Fortymile chapter, in 1895. Fraternal orders like the YOOP and the Arctic Brotherhood played a big part in the social life of the gold rush (University of Alaska Fairbanks Library 70-58-38.)

I drank of this new country and its adventure. It held romance, that kind of romance that does not require the presence of a woman, although a woman would have improved the situation.

Charles Cadwallader on the Iditarod Trail, 1917

What? No health care plan?

The popular A.C. Harris guidebook listed "physical exhaustion, colds, scurvy, rheumatism and snow blindness" as ills the Alaska-Yukon stampeder might face. Some stampeders died of scurvy, pneumonia and typhoid, and a form of meningitis the stampeders called "spinal come-to-Jesus" was the fear of the Klondike trail.

There were few medical supplies and fewer doctors on the trail, so the stampeders carried small "Klondike" medical kits with them. The kits came with turn-of-the-century cures like belladonna plasters, quinine pills, laudanum, toothache drops and paregoric.

Perry Davis Painkiller, a popular item, had other uses besides medicine. Some Fortymile miners used it as a sauce for plum pudding, and trader Jack McQuesten used it as a substitute thermometer. McQuesten set out four vials to gauge the temperature: one each of mercury (quicksilver), whiskey, kerosene and painkiller.

They froze in that order as the temperature plunged. As Arthur Walden remembered, "a man starting on a journey started with a smile at frozen quicksilver, still went at whiskey, hesitated at the kerosene, and dived back into his cabin when the Pain-Killer lay down."

If you love her, build her furniture

A rocking chair carved from a tree trunk at Sunrise, Alaska. Jack Frost gave this chair to his wife Nellie as a Christmas gift in 1897, her first winter in the North. Jack had come to Sunrise in 1896, and Nellie followed a year later (Anchorage Museum 97.2.1, Chris Arend photograph).

A handmade moosehide couch, built by a Kobuk River miner as a gift for his bride-by-mail. The woman he welcomed to his cabin, though, wasn't the woman he had written; she had changed her mind, and her housemate came north to marry him instead (Anchorage Museum 55.6.124, Chris Arend photograph).

The Swedish Queen, a.k.a. Lucille Elliot, a Dawson dance hall favorite (Anchorage Museum B74.1.43).

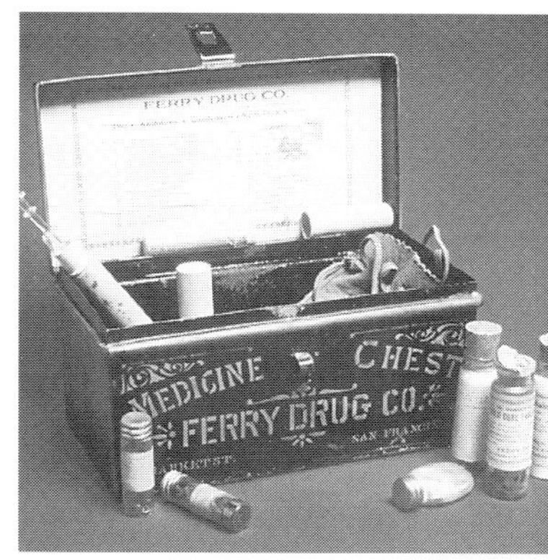

The Klondyke Condensed Medicine Chest, manufactured by the Ferry Drug Company of San Francisco and sold to Klondike stampeders (Alaska State Museum III-0-967, Mark Daughhetee photograph).

The Miners Called Her Mother

In 1894, at age 55, Anna DeGraf crossed Chilkoot Pass in search of her son. He had left their home in Seattle for Alaska, and she had had no word of him for two years. DeGraf walked with a crutch from a broken leg suffered years earlier, but she took her sewing machine with her anyway. As she later wrote, "It takes the trail to find out what kind of stuff is in a person."

She traveled down the Yukon and went to work for Jack McQuesten in Circle, sewing tents and shirts for sale at McQuesten's store. In 1899, she opened a dress shop in Dawson. The miners treated her with respect, calling her Mother, but she never found her son.

Anna DeGraf on the trail to the Yukon (Roger Brown).

Sinking a new prospect hole in the Klondike (Anchorage Museum B74.1.66).

Gold Rush Mining

A Swedish miner who was asked the secret of his luck replied: Just digging holes.

Alaska-Yukon Magazine, March 1907

Over millions of years, ice and water unlocked gold from the crystalline rocks of Alaska and the Yukon. The gold washed down into creeks and rivers and settled out in sand and gravel bars. When the gold rush began, prospectors came into the country, scratching and poking into streambed after streambed. With pick, pan and shovel, they sampled stream gravels, looking for pay dirt.

Where you find it

Most gold rush miners sought placer gold, the gold found in stream deposits. Only in Southeast Alaska was there extensive lode, or vein, mining during the gold rush era. For placer gold, the obvious place to look was a streambed, but it was not always that simple. Some of the richest deposits were deeply buried, and ancient streambeds could be perched high on hillsides. In Nome and other coastal areas, streams carried placer gold to the sea, and ocean waves returned it to land in beach sands.

Placer gold from the Kantishna Mining District, from a claim first worked in 1905 (Louise Gallop, Chris Arend photograph).

Claiming your ground

Once prospectors found gold, they had to measure, mark and file claims to protect the right to mine their discoveries. The claims had names like Six Below Discovery and Two Above, numbered in relation to the "discovery" claim, the spot where the first prospector struck gold on a creek.

Fractions, or small slivers of unclaimed ground, often appeared when surveyors mapped the claims accurately. Prospectors followed the surveyors, ready to pounce on any fraction that the survey cleared. In the Klondike, Dick Lowe's famous fraction, only 89 feet wide, yielded a half-million dollars in gold. Another Klondiker who claimed a tiny sliver of gold ground got his nickname, Nine-inch Jim, from the width of his claim.

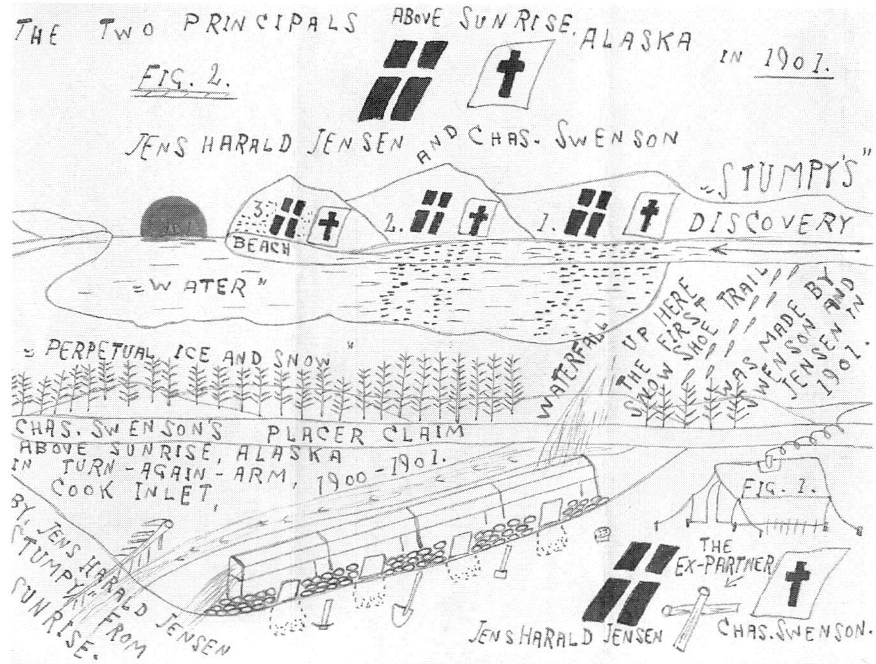

A map to Stumpy's Sunrise claim on the Kenai Peninsula (Anchorage Museum, Alaska Railroad Colonization Collection).

Oh, they scratches the earth and it tumbles out,
More than your hands can hold.
For the hills above and the plains beneath
Are cracking and busting with gold.

From "Label Your Luggage for Klondike," a popular gold rush song.

Basic placer mining

Placer mining is a matter of finding gold-bearing gravels, digging them out of the ground, and separating the heavier gold from the gravel.

Picks and shovels worked well for digging pay dirt out of shallow gravel bars and beaches. Then miners often used a simple, handmade device called a rocker for separating the gold. By adding water to the pay dirt and rocking the device, they washed the material over "riffles" that caught and held the gold.

Using a rocker to separate gold out of pay dirt (University of Alaska Fairbanks Library 89-166-55N).

Down deep: drift mining

In parts of Alaska, gold placers were near enough to the surface to mine as an "open cut." In the Yukon River basin, though, many of the richest gold deposits were deeply buried. To find and mine these pay streaks, miners had to dig deep through the permanently frozen ground.

In the winter miners thawed the earth, inch by inch, with wood fires, sinking a shaft down through the permafrost. If they found a pay streak, the miners dug a drift, or horizontal tunnel, along the trend of the pay dirt.

After digging out the pay dirt, they cranked it to the surface with a windlass and bucket, and stockpiled it in "dumps". Then in summer, when the creeks ran, they washed out the gold.

A weird silence brooded over the wastes of snow.... The flare of red fires parting the twilight marked the beginning of shaft sinking. The snow, the moss, and the fog muffled every foot-fall, deadened every sound. It looked like hell — but it was freezing.

T.A. Rickard in the Klondike, 1897

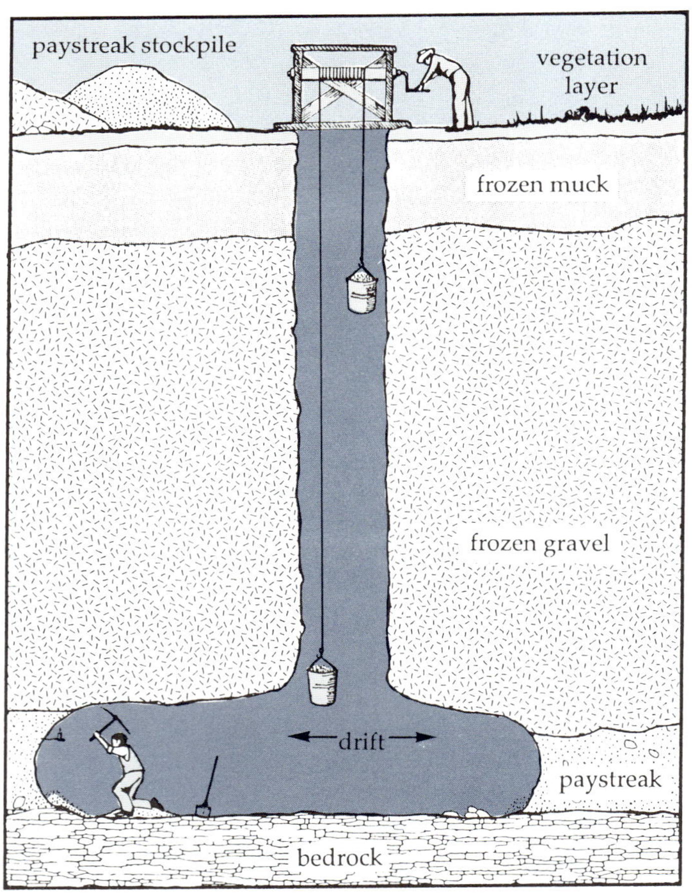

Diagram of a drift mine. When a miner found a pay streak, it was more likely than not just above bedrock, which could be covered by a hundred feet or more of gravel and "muck" (concept source: Parks Canada).

Drift miners used long inclined troughs called sluices to recover the gold. As they ran water down the sluice they shoveled in pay dirt, and the gold settled behind riffles at the bottom of the run.

A miner's candleholder, used to work in drift mines deep below the surface. A shortage of candles could stop a drift miner cold (Anchorage Museum 95.5.5, Chris Arend photograph).

Black sand usually settled out with the fine gold; to separate the two, miners often added mercury. The gold clung to the mercury, and they drove off the mercury with heat from a fire. The earliest miners heated this "amalgam" of mercury and gold in their frying pans.

Industrial mining

After a few years of trial and error, drift miners added some refinements. Hot water and steam replaced open fires for thawing, and cable-run "self-dumpers" eliminated the labor of hand-cranking pay dirt up from a drift. Bigger changes were to come.

Corporate miners bought up blocks of claims and began hydraulic mining, in which water under tremendous pressure blasts away the earth covering a pay streak. They also brought in dredges — huge, floating sluice plants — to work the ground for gold.

Hydraulicking and dredging soon replaced much of the hand work of mining Camp by camp, the work for miners dried up, and the gold rush era faded.

Klondike Kings (and One Queen)

A few of the early Klondike miners became fabulously wealthy. Big Alex McDonald bought a claim for a sack of flour and made millions from it. Charlie Anderson, the "Lucky Swede," was fooled into buying a claim while he was drunk, but he had the last laugh, taking out over a million dollars in gold.

Joe Boyle became a millionaire by winning dredging and hydraulic mining rights in the Klondike from the Canadian government. Boyle came north with $19 in his pocket, as manager of an Australian prize fighter nicknamed the Sydney Cornstalk.

Belinda Mulrooney started her gold rush career by founding two hotels, but soon took over the Gold Run Mining Company and became the richest woman in the Klondike. She made and lost several fortunes in Dawson and Fairbanks.

Top right: Working a sluice near Goldstream Creek, in the Fairbanks district (University of Alaska Fairbanks Library 89-166-17N).

Right: Belinda Mulrooney, gold rush entrepreneur and mine owner, (facing camera, behind sled) with hired packers and loads of goods on the Chilkoot Trail (Alaska State Library 87-682, Winter and Pond photograph).

Below: Yukon Gold Company dredges on Bonanza Creek in the Klondike (Anchorage Museum B64.1.92).

A tent business in Dawson (Anchorage Museum B72.46.181, E.A. Hegg, photograph).

Other Ways to Get Rich

Everybody there turned out to greet us and relieved us of all the money we had, in every way they could think of.

Arthur Walden in Dyea, 1898

Writer Jack London estimated that the Klondike stampeders spent $75 million to get $22 million in gold out of the ground in Dawson's early days. They spent their money on food, tools, tents, stoves, and clothing; on maps and guidebooks; and on steamship and riverboat tickets. Loads of their cash went to Seattle — sales there shot up thirty times during the Klondike rush.

Once stampeders reached the North, they stayed in hotels and roadhouses, bought fresh supplies, and hired packers to carry their goods. They gambled away their cash, and bought whiskey and love.

CONSULT RAYMOND, BOY MEDIUM, about Klondike; truthful predictions, independent slate writing; circles Wednesday, Friday and Sunday evenings. 1422 Second Avenue.

Seattle Times classified ad, February 17, 1898

Hoping to reap huge profits in the Klondike, entrepreneurs packed in magazines, live turkeys, bicycles and even a load of cats to keep Dawson's mice in check. Cattlemen tried at least two cattle drives to Dawson, one from southern British Columbia and one over the Dalton Trail from what is now Haines.

Scam artists and thieves cleaned some stampeders out. In Skagway, Soapy Smith and his gang reigned supreme. In Nome, "chloroform gangs" sneaked into tents, and using chloroform-soaked rags, put sleeping stampeders into such a deep slumber that the thieves could steal everything.

Once just a collection of small, scattered trading posts, the supply business expanded wildly when gold fever spread across the North. The largest suppliers, the Alaska Commercial Company and the North American Transportation and Trading Company, competed tooth and nail, beginning with their trading posts and warehouses on opposite sides of Forty Mile's main street.

More than 100 professional photographers and correspondents, including at least three women, covered the Klondike gold rush. The photos taken by Case and Draper, Winter and Pond, Frank LaRoche, Asahel Curtis, Eric Hegg and other gold rush photographers have become the best-known images of the era.

Soapy Smith, king of the gold rush con men, in his lair in Skagway, 1898 (Anchorage Museum B64.1.45).

Right: A painting of Dawson by Harper's Weekly *and* London Chronicle *correspondent Tappan Adney. Adney spent sixteen months traveling and living with the Klondike stampeders* (Anchorage Museum 94.34.1, Chirs Arend photograph).

A load of twelve tons of liquor and supplies ready to leave Valdez for Fairbanks, 1905 (Anchorage Museum B62.1A.221, P.S. Hunt photograph).

The beach along the coast at Nome has about the same appearance as that of Samoa. In summer the weather is delightful.... The winter in Nome is much less severe than in the northern and northeastern parts of the United States.

Slight exaggerations from a Nome booster booklet, 1905

Edgar Wilson's goat team. Wilson, partner in the trading post at Dyea, used a goat team for packing supplies on the Chilkoot Trail because they were sure-footed and easy to feed (Anchorage Museum B70.35.6).

Brochures intended to boost business for the Nome gold diggings (Keith Christenson, Chris Arend photograph).

Matty Crosby, Iditarod Entrepreneur

Matty Crosby, better known as Miss Tootsie, ran Iditarod's house of prostitution. "The Crosby" was a luxurious establishment complete with running water, lavender glass lamps, lace curtains and flowered wallpaper. She diversified her portfolio by building the town still, the source of Iditarod's most popular drink, and by working mining claims and running a cafe and a boarding kennel for sled dogs.

Crosby came to Dawson just after the Klondike stampede, and eventually worked her way across Alaska to Iditarod. She nearly drowned three times, and she lost sight in one eye after freezing her face driving a dog team. Her glass eye and red wig were her trademarks. She lived in Alaska for over seventy years, not seeing another black for thirty of those years.

Matty Crosby at her diner in Flat (University of Alaska Fairbanks Library P68.69.2344).

Express dog teams from Iditarod in Seward (Anchorage Museum B81.19.2).

By River, Rail and Trail

Iditarod gold team in, 3400 lbs. gold hauled by 46 dogs.

Orville Herning diary, Knik, 1916

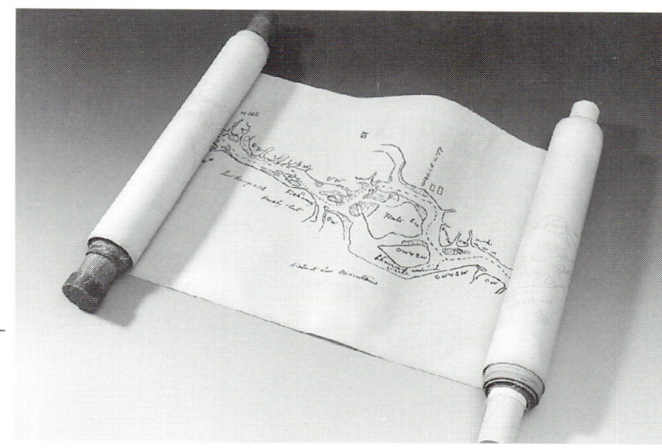

A Yukon River chart. This map of the river, mounted next to the wheel of a steamboat, unrolled as the boat steamed along the river. Yukon navigation was tricky, with many shallows and sandbars that could easily ground a heavily loaded boat (Anchorage Museum B83.21.1, Chris Arend photograph).

Rivers and trails were the first highways in the North. Winter, when the rivers and muskeg were frozen, was the best time for overland travel on foot and by dog team. In summer, boating the rivers was the easiest way to move around in the country. Railroads came later, though not far behind the stampeders.

On the rivers

The gold rush ushered in the heyday of the river steamboat in the North. The 50-foot *Yukon*, dating from the mid-1870s, was the first steamboat on the Yukon River, and one of the few in operation before the Klondike stampede. Among the second generation of Yukon steamboats, the *Alice* and the *Portus B. Weare* carried out the two tons of Yukon gold that electrified the nation in 1897. At the height of the Yukon's steamboat era, about 70 boats worked the river.

The steamboats were the lifelines that enabled the gold camps to develop quickly from rough villages to settled towns, with all the amenities of the outside world.

Traders and merchants needed to plan their shipments carefully, though, because the shipping season was short. In Iditarod, it was very short. Steamboats could reliably navigate the Iditarod River only at high water, in June.

The steamboat Casca *and a stack of firewood, 1908. Steamboats burned great loads of wood cut from the forest* (Anchorage Museum B87.86.16).

The steamboat boom spawned a wood rush that was almost as wild as the gold rush. The boats burned 25 to 40 cords of wood a day, a pile of wood the size of a couple of school buses. White and Native men lived in wood camps on the banks of the rivers, cutting and piling huge stacks of logs for the boats. They denuded the forest as much as three miles back from the Yukon.

Gold rush trails

The trails of the Yukon and Alaska brought many of the stampeders to the gold camps, and as the era unfolded, they became the winter link with the rest of the world. Long trails connected saltwater ports with the camps; the most famous were the Valdez-Fairbanks Trail, now the Richardson Highway, and the Seward-Iditarod Trail, now the Iditarod National Historic Trail.

The snow-covered Iditarod Trail was packed hard and smooth by hikers and dog teams, and an Iditarod traveler could stay at any of 35 roadhouses. There were even a half-dozen places to buy groceries on the trail. Charles Cadwallader hiked the 500 miles to Iditarod carrying only a parka and a change of socks, staying at a new roadhouse every night and eating lunch at several others.

George A. Terrill, an Alaskan sourdough, aged 60, will attempt to mush from Seward to the Iditarod. Terrill froze both feet in the Tanana so that amputation became necessary and now travels on wooden legs.

Seward Weekly Gateway, 1910

At the Pioneer Roadhouse, on the edge of the Alaska Range, guests slept on sheep skins and ate

The sternwheeler Susie *on the Yukon, 1901. The* Susie, Sarah *and* Hannah, *the Alaska Commercial Company's triplet sternwheelers, were the fanciest on the Yukon. Each boat slept 225 passengers and offered staterooms with running water* (Anchorage Museum B90.14.1.201).

Seward-Iditarod Mail Service

MAIL AND PASSENGERS

WEEKLY SCHEDULE BETWEEN SEWARD AND INTERIOR ALASKA

PASSENGER TARIFF NO. 1

All passenger fares must be paid in advance

Seward									
$20	Glacier								
45	$25	Old Knik							
50	30	$5	Knik						
55	35	10	$5	Susitna					
75	55	30	25	$20	Mountain Climbers				
91	71	46	41	36	16	Anderson's			
107	87	62	57	52	32	$16	French Joe's		
154	125	109	104	99	79	63	$35	Takotna	
177	167	142	137	132	112	96	70	$35	Iditarod

Freight and Express Will Be Accepted for Delivery Between Iditarod and Takotna or French Joe's. For Rates See Carriers All Mail Carriers on this route carry an extra sled and team for Passengers, Baggage and Express

GOLD DUST SHIPMENTS — From Iditarod to Seward, $9 per $1,000 in lots less than $10,000; over $10,000, $7.50.

***An advertisement for Harry Revelle's Iditarod Trail dog sled service, 1914, from the* Iditarod Pioneer.**

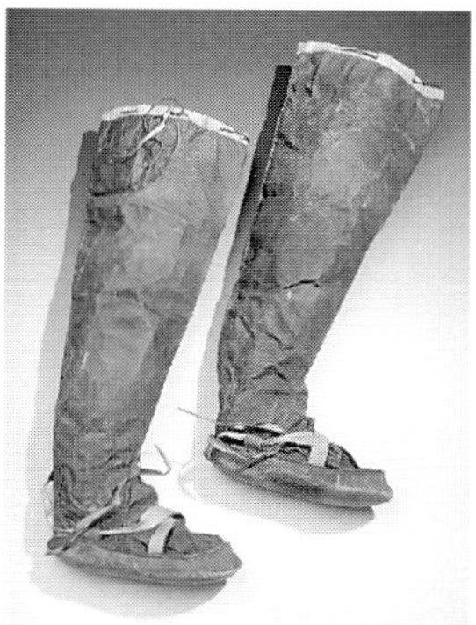

Iditarod miner Arne Erickson's mukluks. Erickson wore his mukluks to hike the snow-covered Iditarod Trail from Iditarod to Seward to visit his sick wife (Anchorage Museum 95.6.1, Chris Arend photograph).

In this 1897 cartoon from Life Magazine, *a gold rush "wheelman" feeds a camp fire with his wooden bicycle frame* (Anchorage Museum, Sherwood Collection).

moose, sheep or caribou steaks. Nearly all of proprietor French Joe Blanchell's provisions came from the land. Once a year, he pulled a sled load of supplies from a landing on the Kuskokwim River, on foot, over 100 miles back to his roadhouse.

The frozen Yukon River was a well-beaten winter trail. A few hardy "wheelmen" even rode bicycles on the ice-covered river, including some who stampeded to Nome from Dawson on two wheels in 1899-1900. One, Ed Jesson, rode into Nome in early 1900, carrying the newspapers that announced to Nome that the U.S. had won the Spanish-American War.

Baldy, Scotty Allan's All-Alaska Sweepstakes champion lead dog, was the inspiration for a popular children's book, Baldy of Nome *(Anchorage Museum B64.1.169).*

Going to the dogs

Dog teams ran mail and hauled supplies to the mining camps in winter. On the Yukon, dog drivers caught chum salmon from the river in summer to feed the dogs that ran the frozen river in winter. Stored in small bales, the dried salmon fed the typical freight team of six or seven heavy Malamute or McKenzie River huskies. The brawny dogs pulled a line of two or three seven-foot sleds with a total load of over a thousand pounds.

Gold rush-era dog breeding and driving reached its pinnacle in Nome. The Nome Kennel Club, founded in 1907, held a series of sled dog races on Nome's trails each winter. The season's climax was the 408-mile All-Alaska Sweepstakes, the first long-distance sled dog race in Alaska. Some of Alaska's most famous drivers of the era — Scotty Allan, John Johnson and Leonhard Seppala — were Sweepstakes champions.

The Camp Comfort Roadhouse on the Valdez-Fairbanks Trail, 1905. Entrepreneurs built roadhouses, offering meals and bunks, about every 20 miles along gold rush trails (Anchorage Museum B62.1A.290, P.S. Hunt photograph).

Laying down tracks

Most gold rush railways were short lines from supply points to the gold diggings. The Tanana Valley Railroad, a 45-mile narrow-gauge railway begun in

1905, connected the Tanana River with the Fairbanks mines. In the Nome district, the first train on Charles Lane's Wild Goose Railroad ran from Anvil Creek to Nome in 1900, just two years after the Anvil Creek gold discovery.

Near the close of the gold rush era, railroad financiers talked big plans, lacing the map of Alaska with proposed railways hundreds of miles long. Almost none of this track was every laid. The exception is today's Alaska Railroad, which was conceived during the gold rush and built after the gold-rush era as a govenment project.

Right top: The little engine that did. A model of a small Porter steam locomotive that saw service in the Klondike and Fairbanks gold districts, and survived the gold rush era to end its career on the Alaska Railroad. The restored locomotive is now on display at Alaskaland in Fairbanks (Tim Sczawinski, Chris Arend photograph).

Right: A locomotive lamp and steam gauge from the Apollo Consolidated Mining Company railroad on Unga Island. Built and delivered in 1897, the Apollo locomotive was one of the earliest in Alaska (Keith Christenson, Chris Arend photograph).

The White Pass and Yukon Route

A chance meeting in a bar led to the building of the White Pass and Yukon Route Railroad, the first major railroad in the North.

At the height of the Klondike stampede, British financier Sir Thomas Tancred traveled to Skagway to check out the feasibility of building a railroad through White Pass to the Yukon. He decided it was impossible, and was ready to leave without a railroad when he ran into Michael (Big Mike) Heney in the St. James Hotel bar. Heney, a Canadian railroad builder, convinced Tancred he could do the job, and the White Pass project was off and running.

Engineers and workmen began work in May of 1898. Working with picks, shovels and blasting powder, they chipped away at canyons and cliff faces and laid the track. Incredibly, they finished the railroad to Lake Bennett by 1899, and to Whitehorse, in the Yukon Territory, by 1900. By then, though, the Klondike rush was history.

A White Pass and Yukon Route train entering Tunnel Mountain (Anchorage Museum B90.14.1.211).

The North-West Mounted Police detachment at Bonanza, in the Klondike (University of Alaska Fairbanks Library 76-92-71).

The Bureaucracy That Worked: Government on the Gold Rush Frontier

I cannot say too much in favor of this wonderful body of men. . . . I am a good American, but I take off my hat to the Canadian North-West Mounted Police as I knew them.

Arthur Walden in the Klondike, 1898

Pioneer explorers and geologists

Army and Geological Survey explorers mapped many of the routes to the gold diggings before the stampedes. Frederick Schwatka led an expedition that accurately mapped the Yukon River in Canada and the Coast Mountain routes to the Klondike for the first time. In 1902, Alfred Brooks explored and mapped the route through the Alaska Range that government crews would blaze as the Iditarod Trail in 1911.

Government geologists mapped the gold districts, giving miners valuable information. Their maps of Nome helped miners trace the ancient beach lines that proved to be some of Alaska's richest paydirt.

With horse flies, gnats and mosquitoes in dense profusion, the Yukon valley is not held up as a paradise to future tourists.

Army explorer Frederick Schwatka
on the Yukon, 1883

Law and order

Before police and judges arrived on the gold rush frontier, the stampeders made their own rough justice through "miners' meetings." A miners' meeting was a people's court where a majority vote decided the verdict. Anyone could call a meeting to settle a dispute or accuse someone of a crime. Convicted thieves usually received a strong suggestion to leave the country.

In Circle, the Circle City Miners Association wrote bylaws, kept minutes of its meetings, and settled all kinds of disputes. In one case, a woman who had borne a dancehall fiddler's child asked the miners to enforce the fiddler's promise to marry her. In just a few minutes of deliberation, the association ordered the fiddler to pay the woman's maternity expenses and marry her by 5:00 that afternoon.

The first official arm of the law on the new fron-

A miners' meeting called to try camp thieves at Sheep Camp, on the Chilkoot Trail (Museum of History and Industry 16454).

The Allen Expedition: A Thousand Miles on an Empty Stomach

In 1885, a three-man backpack expedition led by Lieutenant Henry Allen traced a swath across Alaska from Prince William Sound to the Bering Sea. Their 20-week journey would result in the first accurate maps of three major rivers: the Copper, Tanana and Koyukuk.

Never very well provisioned, they hit a low point on April 13, Allen's birthday, when, as Private Fred Fickett wrote in his journal, they "celebrated . . . by eating rotten moose meat." Four days later, though, Fickett allowed that rotten moose meat would now be "a delicacy". When they ate well, dinner was rabbit in juice thickened with a handful of flour.

They survived the journey with the help of Native guides and villagers. Back in Washington, they published the first maps of the three valleys, each of which would be poked and prodded for gold by prospectors later in the era.

Henry Allen, center, with Fred Fickett, left, and Cady Robertson, right, at the end of their Copper-Tanana-Koyukuk expedition of 1885 (University of Alaska Fairbanks Library 72-164-1N).

Fort Liscum, near Valdez, a U.S. Army post established in 1900 (Anchorage Museum B62.1.1050).

tier was the North-West Mounted Police. A detachment of Mounties under Inspector Charles Constantine set up a station at Forty Mile in 1895, and ended the miners' meeting in Canada. After the turn of the century, the miners' meetings in Alaska began to fade into history.

The second gold rush assignment for the North-West Mounted Police was to set up customs stations at Chilkoot Pass and White Pass. A small group of Mounties arrived at the passes early in 1898, at the height of the Klondike stampede. As the thousands of stampeders filed past, the Mounties checked them for the required year's worth of supplies and collected customs duties. At Lake Lindeman, in what the Canadians regarded as their territory, they evicted two American businessmen who had set up a warehouse and townsite. By setting up their customs stations in the passes, the Mounties had established the U.S./Canada border.

In Alaska, the Army brought the first official law and order. Federal courts followed, led by Judge James Wickersham, who established a court at Eagle and then traveled to Nome to clean up a mining claim mess after a riot of claim jumping.

Crooks of all kinds plagued Skagway and Nome in their very early days, but other gold rush towns - Dawson, with the Mounties in charge, and Iditarod, which never even had a jail - had little crime.

Mail, trails and telegraph lines

Neither grizzlies nor ground blizzards could halt the mail, the stampeders' only link to home and family. The Army built Alaska's mail trails, and its wagon roads and telegraph lines as well. Between 1905 and 1912, the Army's Alaska Road Commission built 2,000 miles of trails, 800 miles of wagon roads, and 600 miles of winter sled roads. The Signal Corps set up 1,500 miles of telegraph cable between 1900 and 1903

The arrival of the mail at Tagish Post, on the Yukon, in 1898 (Museum of History and Industry 12371).

A Han family camp near Dawson, 1898 (Anchorage Museum B74.1.6).

Native Peoples and the Gold Rush

When Skookum Jim found gold, that's the time everything changed.

Yukon Native elder Annie Ned, 1975

Robert Service wrote in his famous poem *The Spell of the Yukon* that the Yukon was "a land where the mountains are nameless, and the rivers all run God knows where." Actually, the Yukon and Alaska Native people already knew fairly well where the rivers ran when the stampeders arrived. The Native populations were not large, but the land was not empty.

Native men selling moose meat in Dawson (Anchorage Museum B74.1.9).

Changing times

Most Alaska and Yukon Native groups had already had contact with white fur traders, prospectors, or whalers before the gold rush era. The crowds that came with the gold rushes, though, dwarfed the numbers of whites in the country before the stampedes. Between 1880 and 1900, the white population of Alaska grew from 400 to over 30,000, surpassing the Native population by the turn of the century.

The gold rush caused major changes in the Native cultures directly in the path of the stampedes — the Tlingit of Southeast Alaska, the Athabaskan

Chief Isaac, left, the chief contractor for packing supplies over Chilkoot Pass in the early days of the Klondike rush (Anchorage Museum Archives Collection).

peoples of the Interior, and the Inupiat of northwest Alaska. The stampeders delivered alcohol and illness, disrupted hunting and fishing, and set in motion the loss of land that disintegrated Native ways of life.

On the other hand, the gold rush also brought new tools and goods and a chance to profit from the stampeders. Natives worked as packers, guides, wood cutters, riverboat workers and pilots, and market hunters. Some made and sold Native art to the stampeders. An Ahtna man, Doc Billum, ran a ferry on the Copper River.

The Tlingit business empire

Before the gold rush era, the Tlingit people at the head of Southeast Alaska's Inside Passage treated the passes through the Coast Mountains as private property. The passes were important

A Fish Camp on the Klondike

When explorer J.E. Spurr passed the Trondiuck River in 1896, Han men were spearing salmon in the river from birchbark canoes. The river's name meant "hammer water," from the stakes hammered into the riverbed to trap the fish. Where the Trondiuck meets the Yukon, Spurr reported a camp of 200 Han. The Han dried heaps of salmon for the Yukon winter, and the Trondiuck was a choice fishing stream.

The same summer Spurr passed the Trondiuck camp, the Mason-Carmack party discovered gold just a few miles away. Soon the river had a new name, the Klondike, and by the next summer, there was a new town named Dawson at the mouth of the river.

The Han fish camp at the mouth of the Klondike River c. 1895, before the Klondike discovery (Yukon Archives 82-358).

trade routes, and the Tlingit closed them to maintain a trade monopoly with the Yukon people.

They bartered their own products — fish, blankets, wooden baskets and boxes — for hides, furs, and skin clothing from the Yukon. The Tlingit took their trading seriously; in 1852, they burned the Hudson's Bay Company post on the Yukon at Fort Selkirk to eliminate the British competition. When white prospectors demanded to use Chilkoot Pass to prospect for gold in the Yukon, the Tlingit agreed on the condition that the prospectors have nothing to do with the fur trade.

During the 1880s and 1890s, the Tlingit held a monopoly on packing supplies over the pass. Travelers crossing the pass paid the packers up to a dollar a pound for carrying their gear. Women and young boys joined the Tlingit men in carrying large loads bundled together with rope and supported with just shoulder straps and a head strap.

When the frenzy of the 1897-98 Klondike rush reached its height, white men brought in pack animals and built tram lines to compete with the Tlingit packers, and the monopoly dissolved.

It seemed marvelous beyond measure how these small Indians, not averaging, I believe, over 140 pounds each, could carry 100 pounds up such a precipitous mountain.

Frederick Schwatka with a group of Tlingit packers at Chilkoot Pass, 1883

Finding gold

The discoveries of several major gold fields involved Native people. Kowee, a Tlingit leader of the Auke Bay band, chaperoned two white prospectors to the discovery of the Juneau Gold Belt. Two Russian-Native men found the gold that established Circle, and two of the three prospectors in the party that struck gold in the Klondike were Tagish, the Native people of the upper Yukon.

In Alaska, Natives could not stake mining claims because U.S. law did not recognize them as citizens. In the Yukon, a few Natives staked and mined claims. Skookum Jim and his nephew Dawson Charlie, both members of the Klondike discovery party, staked claims that made them rich. Jim and Charlie started a second, minor gold rush in 1903 when they found gold near Kluane Lake in the Yukon.

Skookum Jim Mason's claim, No. 1 Above Discovery on Bonanza Creek. Mason, a Tagish man, discovered the gold that brought on the Klondike stampede (Anchorage Museum B90.14.1.194).

Chilkoot Pass, 1996, with collapsible boats left near the pass during the Klondike gold rush (Dean Littlepage photograph).

A Century Later: The Legacy of the Gold Rush Era

My deceased husband's grandfather, Charles E. Packard, arrived in Dyea in August (1897). . . . Charles brought his 16-yr. old son, Wilbert, who contracted meningitis and died in Dyea, Easter Sunday April, 1898. . . . I came over the Chilkoot in memory of all of them.

Chilkoot hiker Kathleen Packard's entry in the Lake Lindeman log book, 1995

The North's gold

The great gold fields of the gold rush era were the Klondike, Fairbanks, Juneau, and Nome. A hundred years after the Klondike stampede, miners are still working these and other districts with new technology and methods. Most gold mining is still placer mining, and gold is still the mineral most prospectors seek.

The first gold pour at the Fort Knox Mine outside Fairbanks, 1996. The molten gold cools into gold bars. Fort Knox is the newest and largest of Alaska's gold mines (Fairbanks Gold Mining Company).

Modern gold miners have never surpassed what the gold rush miners did in 1906, when they dug more than a million ounces of gold from the earth. The largest nugget found during the gold rush is still the Alaska champion, a seven-pound chunk of gold found in Nome's Anvil Creek in 1901.

All the gold mined in Alaska since the beginning of the gold rush would fit into an average-sized living room, but it would weigh over 1,000 tons.

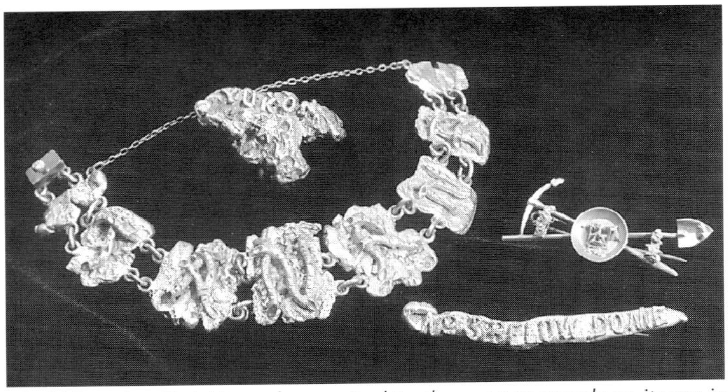

Gold jewelry. Gold is no longer tied to the money supply as it was in the 1890s, and most gold is now used in making jewelry. Its other major use is in electronics (Lillian S. Watson Collection, Pioneers of Alaska Auxiliary #4, Chris Arend photograph).

Piecing together gold rush history

The Alaska-Yukon gold stampeders left a rich legacy of diaries, stories and photographs. Archaeologists have been able to add to our understanding of life in the early mining camps by exploring and excavating some of the sites.

Archaeologists working at the "Barnette Site," the spot where E.T. Barnette set up the trading post that became Fairbanks, have recently uncovered 35,000 objects from early life in the city. The objects came from the remains of a saloon, a dock, a warehouse and a home. One of the important things the archaeologists learned from the study is how quickly Fairbanks changed from a frontier camp to a settled city with all the amenities of the outside world.

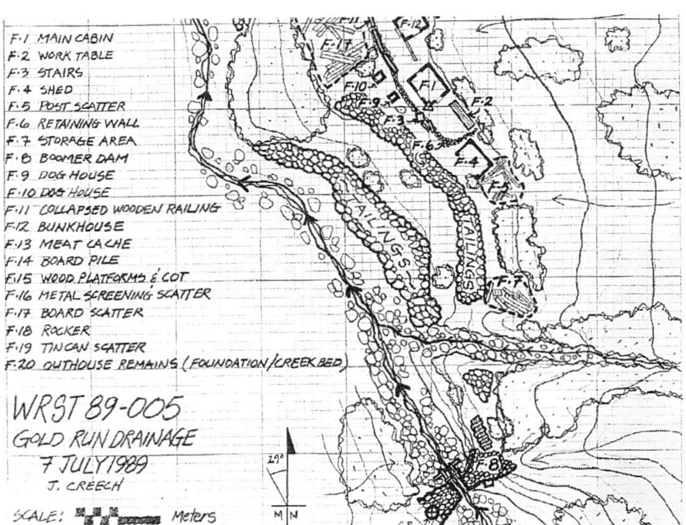

An archaeologist's gold rush site map from the Wrangell Mountains. Archaeological work combined with written accounts from the era have helped students of the gold rush piece together what everyday life was like in the mining camps (National Park Service).

Popular culture

The Yukon and Alaska gold rushes fascinated the United States and the world. The exotic North made its way into the popular imagination through fiction, poetry, songs, movies, games, and commercial products.

Odd logic connected products like Klondike Head Rub and Klondike Gold chewing gum to the gold rush. Tunes like "The Chilkoot March" made their debuts as sheet music while the stampeders were still plodding up the trail. Gold rush-inspired pulp fiction for young people starred heroes like Diamond Dick Junior and Young Klondike.

Jack London, who came to the Klondike from San Francisco at the age of 21, is the best-known prose writer from the era. His short story "To Build a Fire" and his novels *The Call of the Wild* and *White Fang* still carry their original appeal. Robert Service, who came to the Yukon six years after the

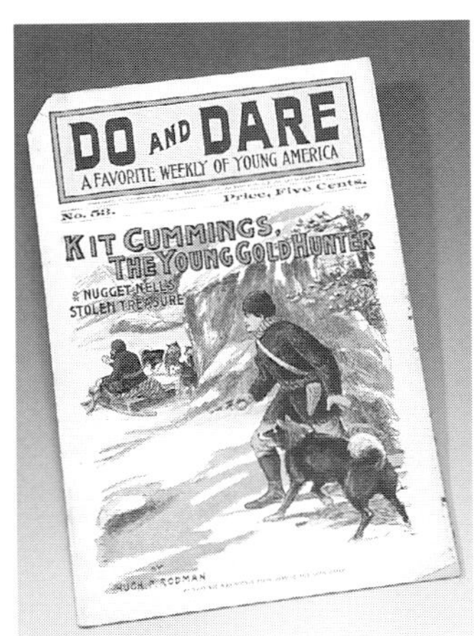

Left: Do and Dare, *gold rush pulp fiction for young people, published in 1901* (Anchorage Museum, Sherwood Collection, Chris Arend photograph).

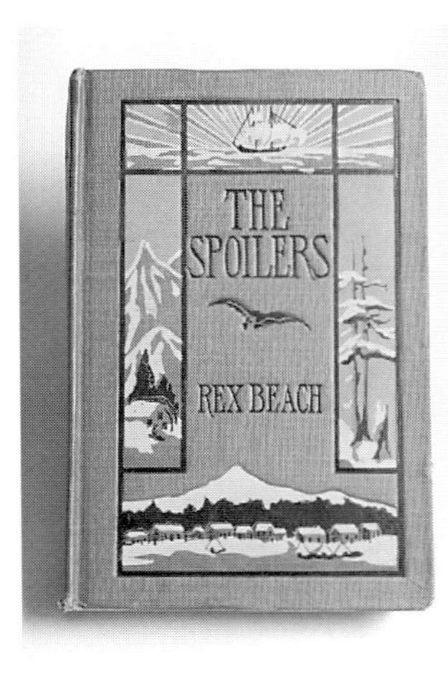

Right: The Spoilers, *a Rex Beach novel based on the McKenzie-Noyes claim-jumping conspiracy during the Nome gold rush* (Anchorage Museum Archives Collection).

gold rush, wrote the haunting poem "The Spell of the Yukon" and scores of other gold rush poems.

I wanted the gold, and I sought it;
I scrabbled and mucked like a slave.
Was it famine or scurvy — I fought it;
I hurled my youth into a grave.

The opening lines from Robert Service's
The Spell of the Yukon

Gold rush places today

Many modern cities in Alaska and the Yukon — Dawson, Whitehorse, Nome, Fairbanks, Seward, Valdez, Eagle, Circle, Juneau, Skagway, Wrangell and others — began as gold rush-era settlements. The Alaska Railroad, the White Pass and Yukon Route Railroad, and the Seward and Richardson Highways all began as gold rush routes. Steamboats played a role on the Yukon River until the 1950s; now barges deliver freight to the villages on the river. The Iditarod Trail near Anchorage is now a hiking trail, and dogs race on the trail between Anchorage and Nome every year in the Iditarod Sled Dog Race.

The ghost town of Iditarod (Dean Littlepage photograph).

Hollywood has told and retold its version of the gold rush era through the years (Candy Waugaman, Chris Arend photograph).

A raft built by two German adventurers on the Yukon at Circle (Dean Littlepage photograph).

Along Bonanza Creek in the Klondike, where it all began in 1896, modern prospectors can buy claims from a real estate chain. (Dean Littlepage photograph).

Further Reading

The following list is a fraction of the rich and wonderful literature of the gold rush era. The complete list of sources consulted for *Gold Fever in the North* includes these titles but is too long to include in its entirety here.

First-hand accounts:

Tappan Adney
The Klondike Stampede. University of British Columbia Press, 1994 (originally published 1900).

Laura Berton
I Married the Klondike. McClelland & Stewart, 1961.

Anna DeGraf
Pioneering on the Yukon, 1892-1917. Archon Books, 1992 (Roger S. Brown, ed.).

Albert W. Morgan
Memories of Old Sunrise: Gold Mining on Alaska's Turnagain Arm. Cook Inlet Historical Society, 1994 (Rolfe G. Buzzell, ed.)

First-hand accounts, out of print:

Kenneth Gideon
Wandering Boy. Merkle Press, 1967.

Frederick Schwatka
Along Alaska's Great River. Alaska Northwest Books, 1983 (originally published 1885).

Arthur T. Walden
A Dog-Puncher on the Yukon. Houghton Mifflin, 1928.

James Wickersham
Old Yukon Tales, Trails and Trials. Washington Law Book Company, 1938.

Secondary works:

Francis Backhouse
Women of the Klondike. Whitecap Books, 1995.

Pierre Berton
The Klondike Fever: The Life and Death of the Last Great Gold Rush. Alfred A. Knopf, 1984.

Terrence Cole
Crooked Past: The History of a Frontier Mining Camp: Fairbanks, Alaska. University of Alaska Press, 1991.

Terrence Cole
Nome: City of the Golden Beaches. Alaska Geographic Society, 1984.

Michael Gates
Gold at Fortymile Creek: Early Days in the Yukon. University of British Columbia Press, 1994.

Jane Haigh/Clair Murphy
Gold Rush Women. Alaska Northwest Books, 1997.

William R. Hunt
North of 53°: The Wild Mining Days of the Alaska-Yukon Mining Frontier, 1870-1914. Macmillan, 1974.

David Neufeld/Frank Norris
Chilkoot Trail: Heritage Route to the Yukon. Lost Moose, 1996

Archie Satterfield
Klondike Park: From Seattle to Dawson City. Fulcrum, 1993.

Pictorial histories:

Murray Morgan
One Man's Gold Rush: A Klondike Album. University of Washington Press, 1967 (photos by E.A. Hegg).

Jasper Wyman
Journey to the Koyukuk: The Photos of J.N. Wyman, 1898-99. Pictorial Histories, 1988.

Sources of Quotations

Inside cover	Ralph Lomen Collection, University of Alaska Fairbanks Archives.
Page 1	Charles Judge, *An American Missionary*. Ossining, 1907.
Page 3	Josiah Spurr, *Through the Yukon Gold Diggings*. Eastern, 1900.
Page 5 left	John Sydney Webb, "The River Trip to the Klondike," in *The Century Magazine*, March 1908.
Page 5 right	*Seattle Times* (Seattle, WA), July 24, 1897.
Page 6	Quoted in Pierre Berton, *The Klondike Fever*. Alfred A. Knopf, 1984.
Page 7	Louie B. May diary, Alaska State Library, Juneau.
Page 8	Fred Crewe, *Poems of Klondyke's Early Days and Alaska's Long White Trail*. North American Press, 1921.
Page 9	Emily Craig, *A Pioneer Woman in Alaska*. Caxton, 1948.
Page 10	Kenneth Gideon, *Wandering Boy*. Merkle Press, 1967.
Page 12 left	*Yukon Sun* (Dawson City, Yukon Territory), January 17, 1903.
Page 12 right	Anton Eide, Unpublished report to the Alaska Road Commission, August 18, 1910.
Page 14	Laura Berton, *I Married the Klondike*. McClelland & Stewart, 1961.
Page 15	Harold and Zora Peckenpaugh, *Nuggets and Beans*. Carlton, 1973.
Page 16	Charles Cadwallader, *Reminiscences of the Iditarod Trail*. n.p., n.d.
Page 18	"Sourdough," "Mining the Northland," *Alaska-Yukon Magazine*, March 1907.
Page 19	Paul Roseland Collection, Anchorage.
Page 20	T.A. Rickard, "Mining Methods in the North - I." *Mining and Scientific Press*, December 12, 1908.
Page 22 left	Arthur T. Walden, *A Dog-Puncher on the Yukon*. Houghton Mifflin, 1928.
Page 22 right	*Seattle Times* (Seattle, WA), February 17, 1898.
Page 24	"Nome: A Wonderland of Wealth," 1905, Al Polet Collection, University of Alaska Fairbanks Rasmuson Library.
Page 25	Orville Herning diary, Herning Collection, Anchorage Museum.
Page 26	*Seward Weekly Gateway* (Seward, AK), January 22, 1910.
Page 29	Arthur T. Walden, A Dog-Puncher on the Yukon. Houghton Mifflin, 1928.
Page 30	Frederick Schwatka, *Along Alaska's Great River*. Alaska Northwest Books, 1983 (first published 1885).
Page 32	Rab Wilkie and the Skookum Jim Friendship Center, Skookum Jim. Yukon Tourism, Heritage Branch, 1992.
Page 34	Frederick Schwatka, *Report of a Military Reconnaissance in Alaska, Made in 1883*. Government Printing Office, 1885.
Page 35	Chilkoot Trail register, Klondike Gold Rush National Historic Park Collection, quoted in David Neufeld and Frank Norris, Chilkoot Trail. Lost Moose, 1996.
Page 37	Robert Service, *The Spell of the Yukon*, first published in 1907.
Back cover	Tappan Adney, *The Klondike Stampede*. University of British Columbia Press, 1994.

A Gold Rush Timeline

1914 Jay Livengood finds gold on Livengood Creek north of Fairbanks.
1913 Billy James and Nels Nelson discover gold at Chisana in the Wrangell Mountains.

1911 The Alaska Road Commission blazes the Iditarod Trail.

1908 John Beaton and William Dikeman strike gold on the Iditarod River.

1903 Miners from Dawson stage a -50° stampede to the Tanana.
1902 Felix Pedro's discovery near the Tanana River leads to the founding of Fairbanks.

1899 Stampeders begin the rush to Nome.
1898 30,000 stampeders reach the Klondike. "The 3 lucky Swedes" strike gold at Anvil Creek (Nome).
1897 The Klondike stampede begins.
1896 The Carmack/Mason party strikes gold in the Klondike. Prospectors rush to Hope and Sunrise on the Kenai Peninsula.

1893 The Panic of 1893 plunges the U.S. into economic depression.
1892 A discovery on Birch Creek opens the Circle Mining District.

1888 Al King rediscovers gold on the Kenai Peninsula.

1886 Howard Franklin and Henry Madison strike gold on the Fortymile River in interior Alaska near the Canadian border.

1880 Kowee, Richard Harris and Joe Juneau discover the Juneau Gold Belt. Tlingit Natives allow prospectors to cross Chilkoot Pass.

1873 Jack McQuesten, prospector and trader, reaches the Yukon River.
1872 Prospectors make a second Stikine gold strike.

1867 U.S. purchases Alaska from Russia.

1861 Buck Choquette finds gold on the Stikine River in British Columbia near Wrangell.

1849 Russian engineer Peter Doroshin discovers gold on the Kenai Peninsula.